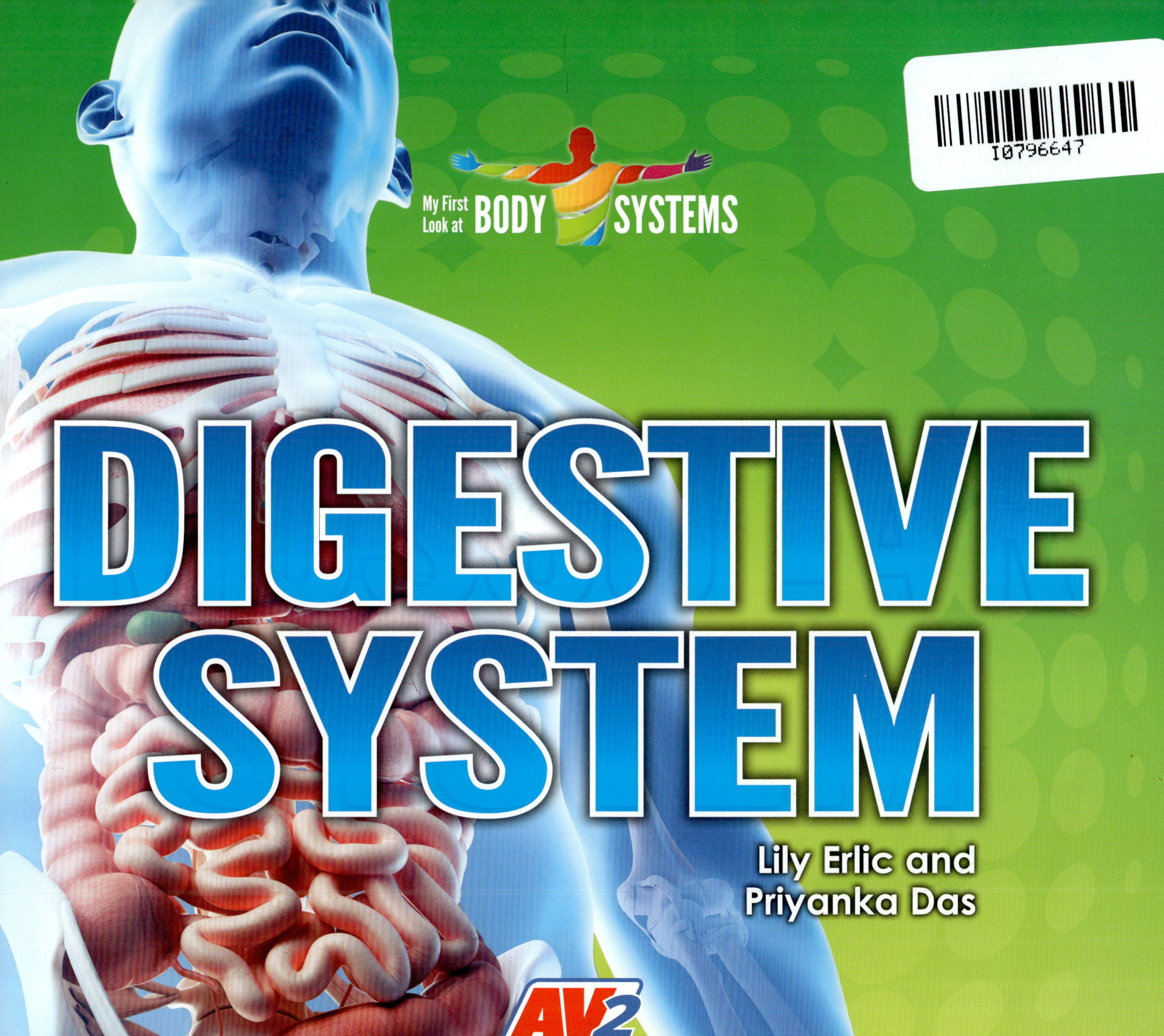

My First Look at BODY SYSTEMS

DIGESTIVE SYSTEM

Lily Erlic and
Priyanka Das

AV2
www.av2books.com

Step 1
Go to **www.av2books.com**

Step 2
Enter this unique code
RITLQCR97

Step 3
Explore your interactive eBook!

My First Look at BODY SYSTEMS

DIGESTIVE SYSTEM

Start!

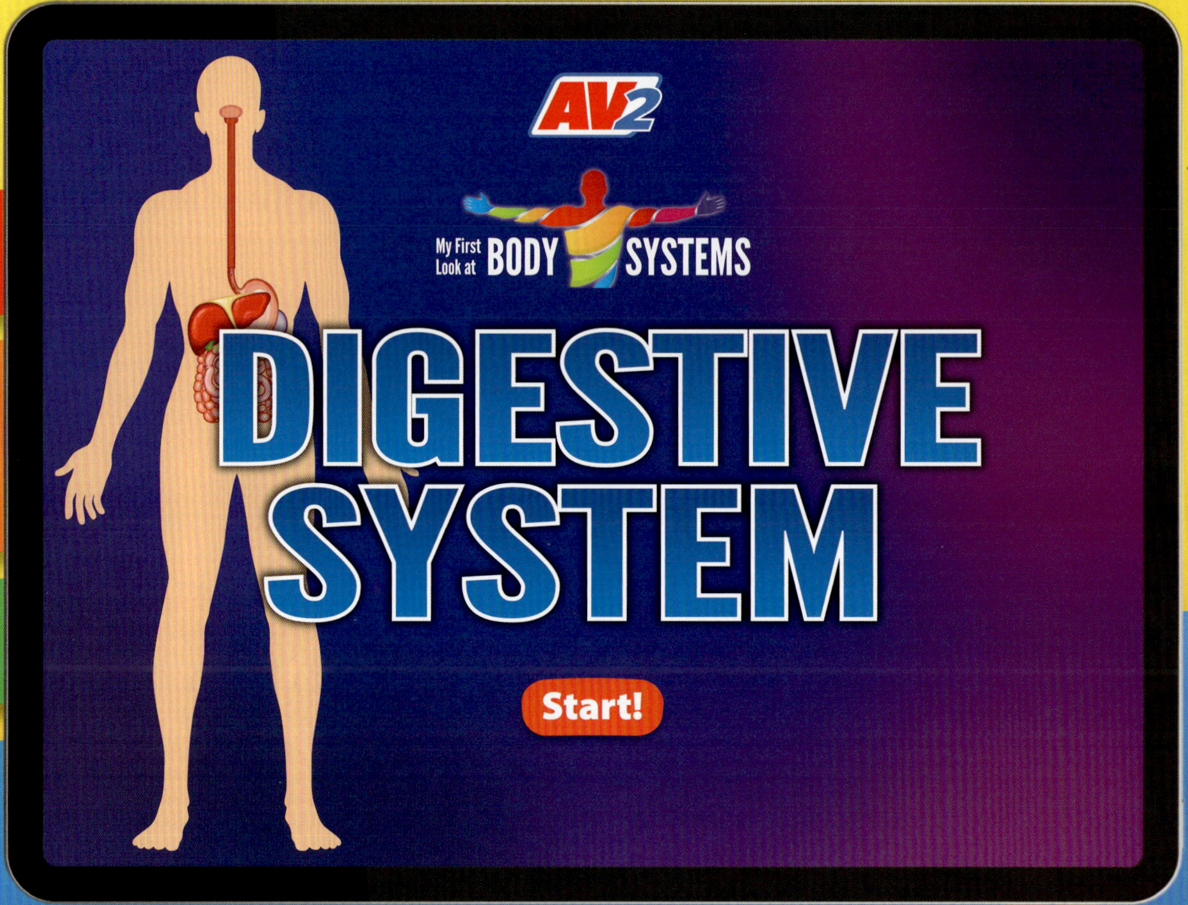

AV2 is optimized for use on any device

Your interactive eBook comes with...

Audio
Listen to the entire book read aloud

Videos
Watch informative video clips

Weblinks
Gain additional information for research

Try This!
Complete activities and hands-on experiments

Key Words
Study vocabulary, and complete a matching word activity

Quizzes
Test your knowledge

Slideshows
View images and captions

CONTENTS

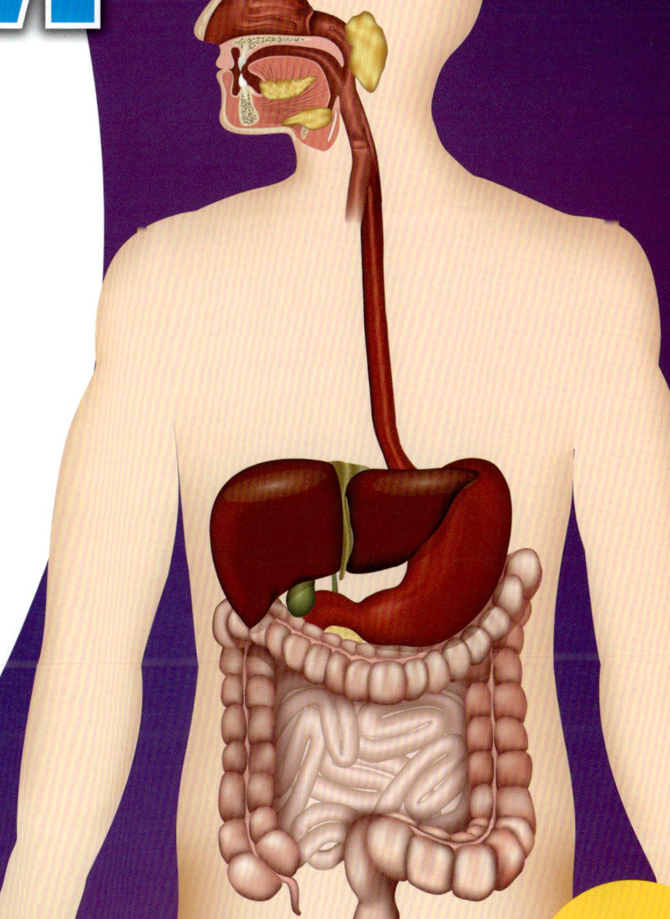

Digestive System

Do you remember what you had for breakfast this morning? The food you ate is still in your body. **It is being used to give you energy.**

The body needs nutrients to work properly. It gets these nutrients from food and water. **The digestive system helps the body break down food.** Every time you eat or drink something, it goes through your digestive system.

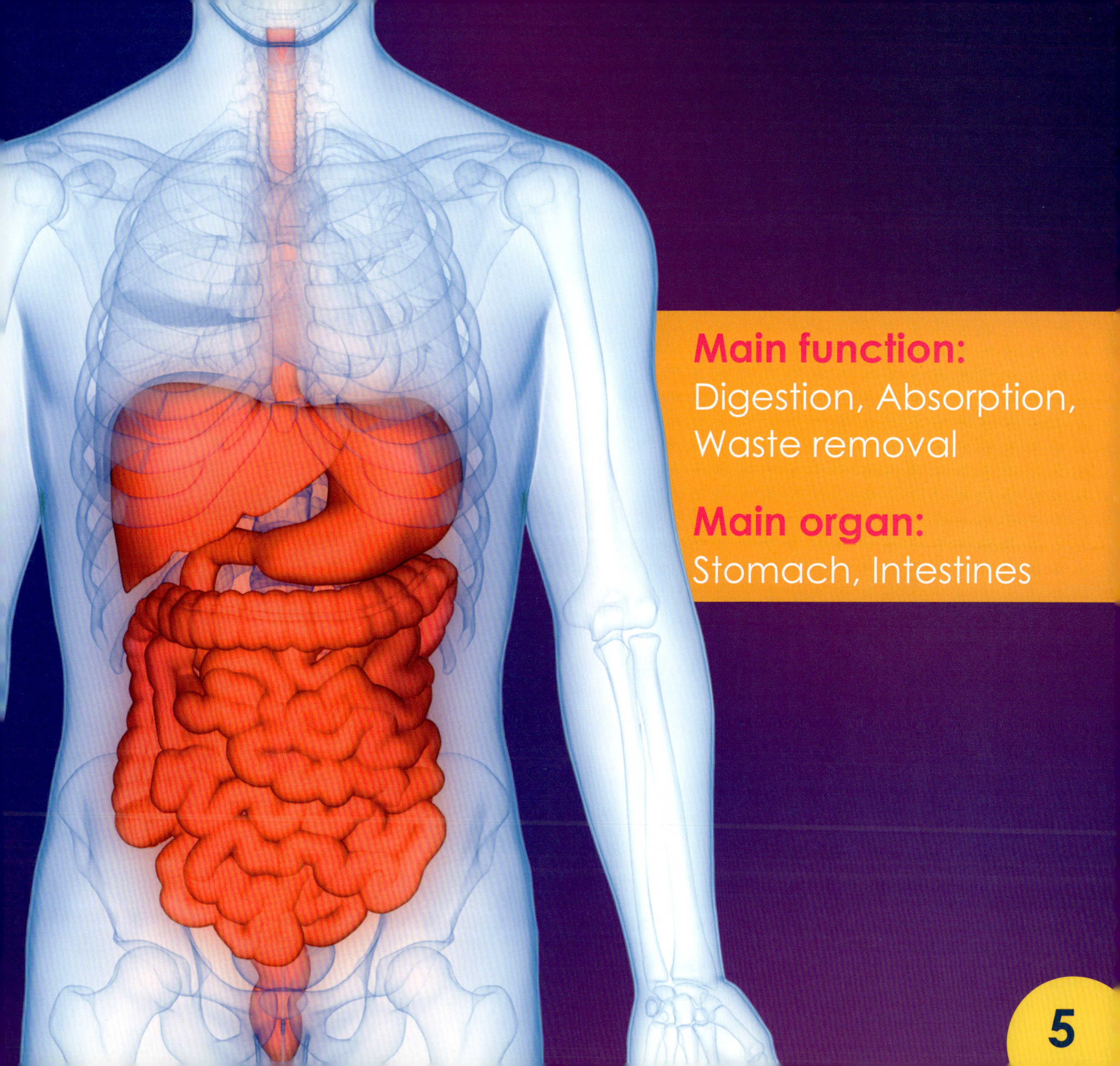
Main function:
Digestion, Absorption,
Waste removal
Main organ:
Stomach, Intestines

All about the Digestive Tract

The digestive tract is part of the digestive system. Food enters the body through the mouth and moves down the digestive tract. **The digestive tract breaks food into small parts that the body can absorb.** This process is called digestion.

There are many nutrients in food. **The body absorbs all the nutrients that it can.** Some parts of food that the body cannot use are left over. This waste is removed from the body.

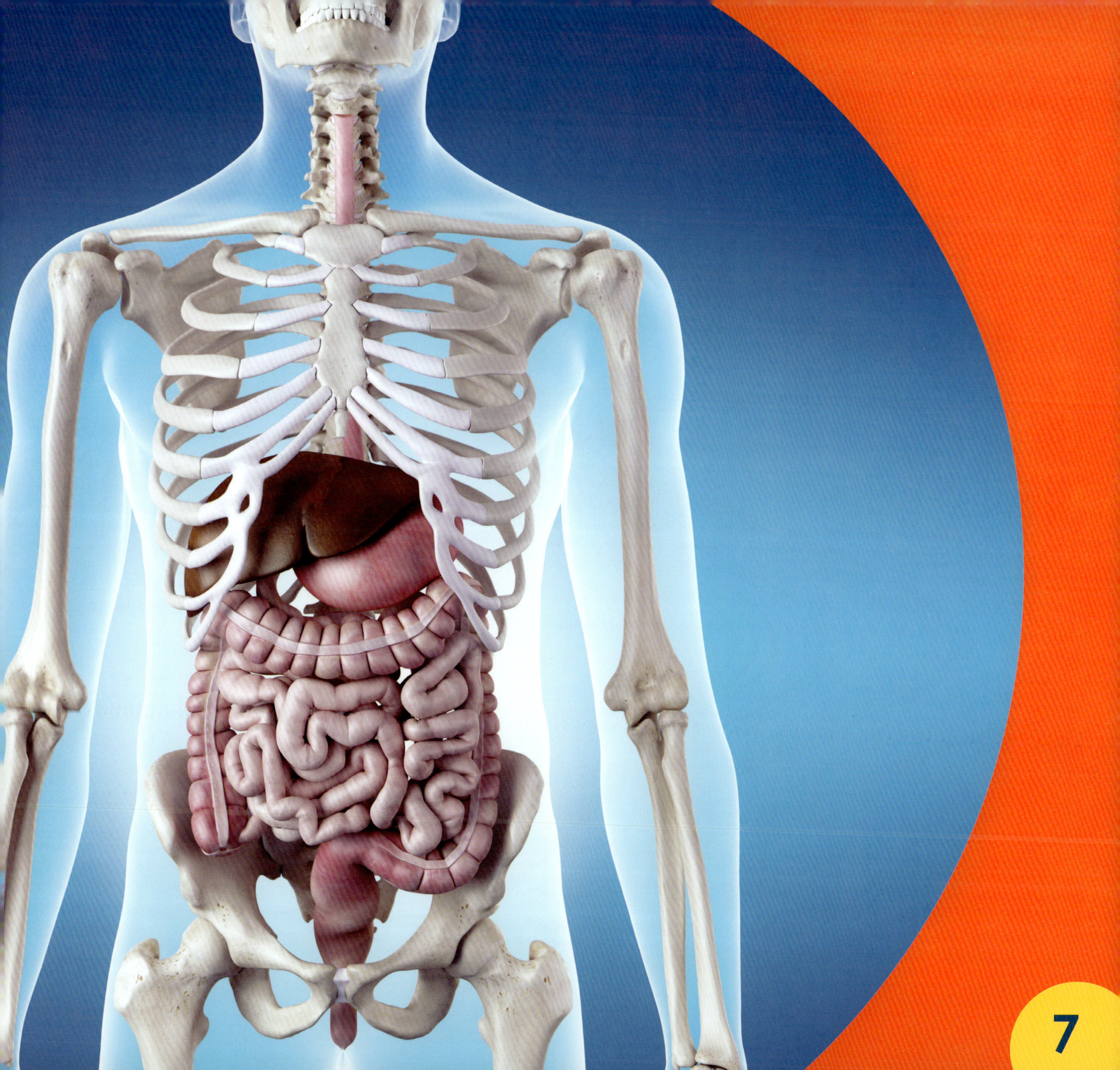

Parts of the Digestive System

The digestive tract is a long, twisting tube with many organs. It runs from the mouth to the anus. The organs of the digestive system work together to break down and absorb food.

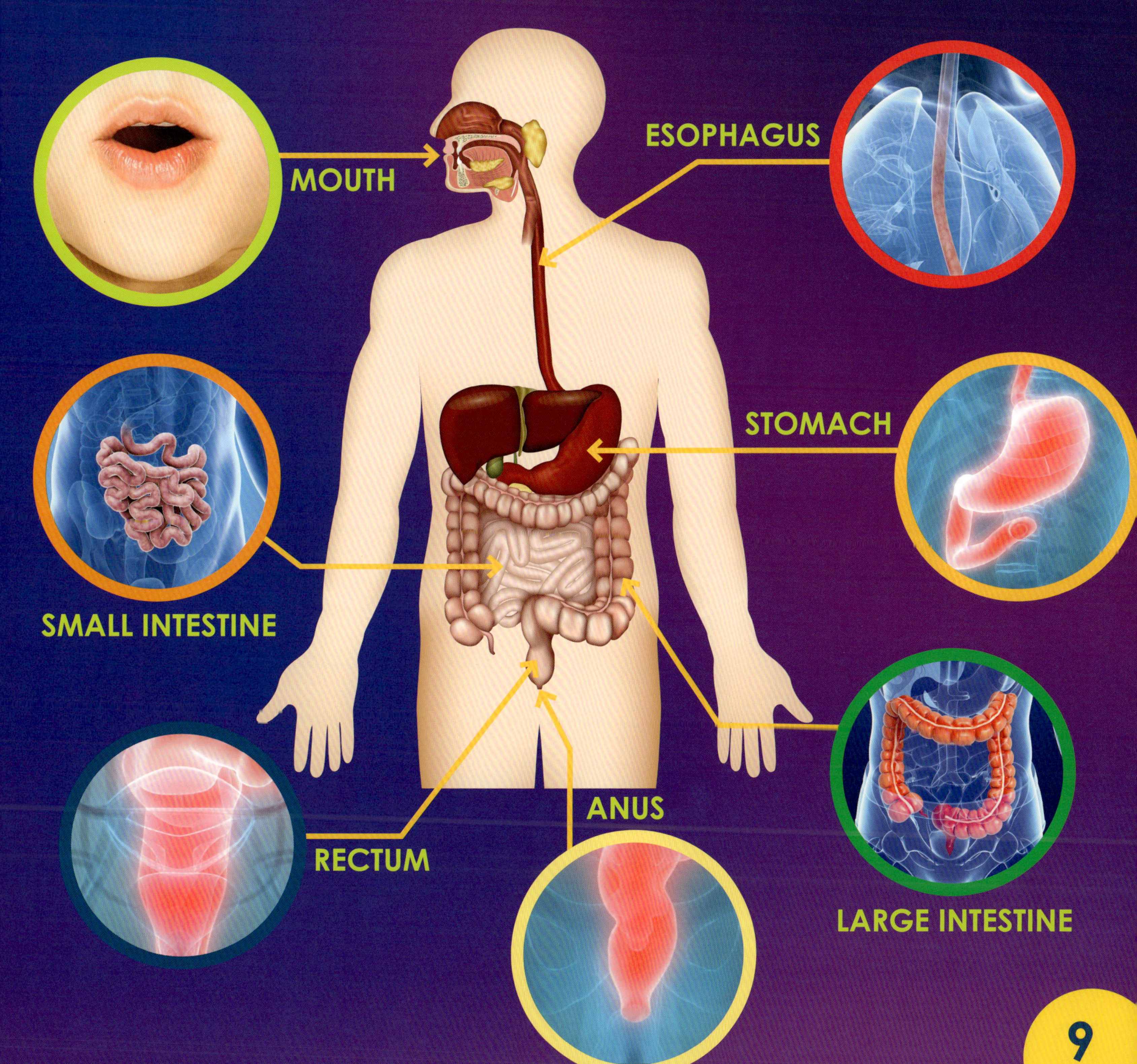
MOUTH
ESOPHAGUS
STOMACH
SMALL INTESTINE
LARGE INTESTINE
RECTUM
ANUS

Mouth

Digestion begins before you even take a bite of your food. It starts in the mouth. When you smell food, your mouth makes a liquid called saliva. Saliva helps break down food. As you eat your meal, your mouth makes more saliva.

Your mouth moves when it chews food. **Teeth grind the food into smaller parts.** Then, your tongue helps you swallow the food.

Babies are born with **20 teeth** in their mouths. Adults have **32 teeth**.

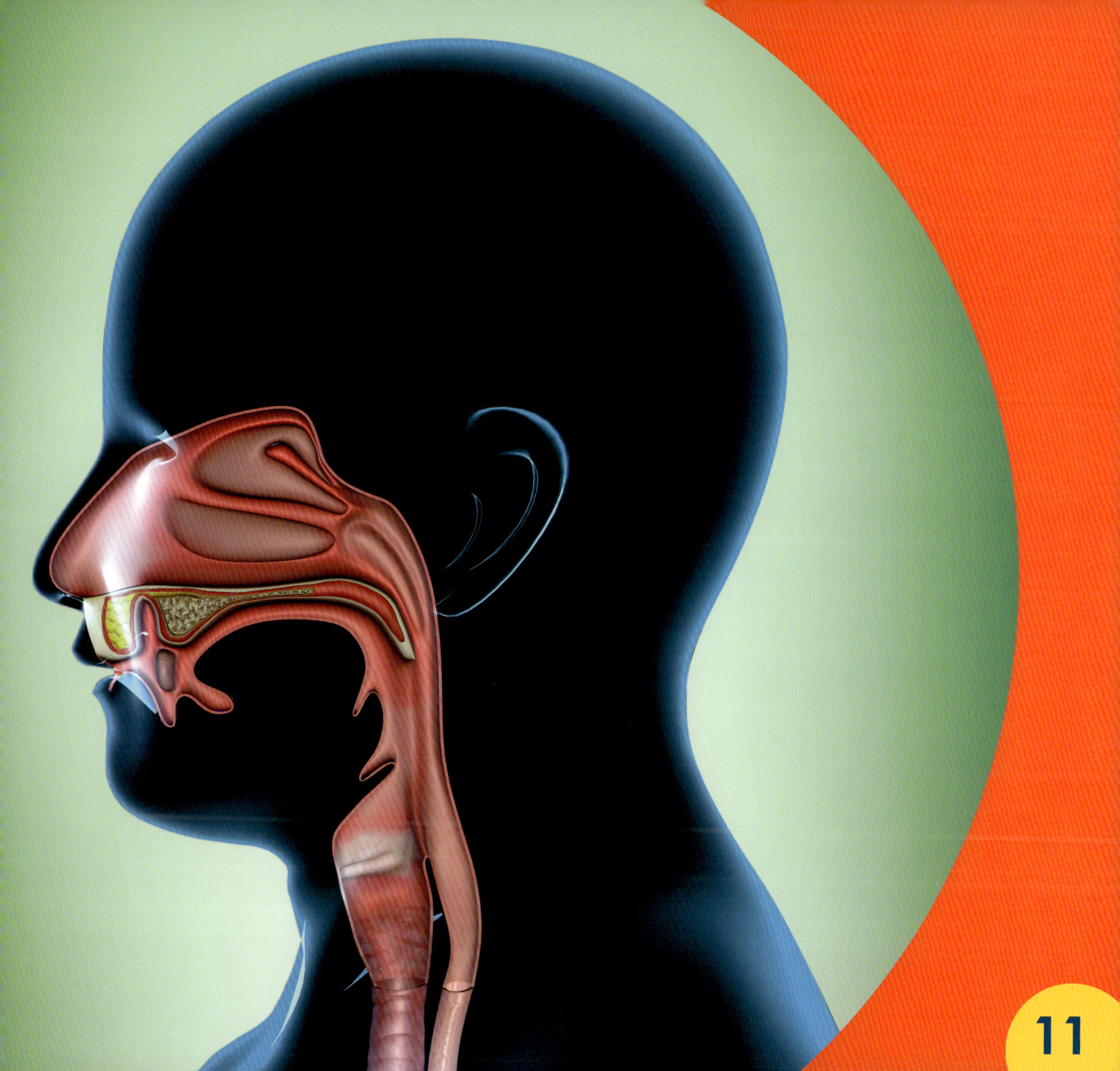

Esophagus

Food that is swallowed is pushed down to the esophagus. The esophagus is also called the food pipe. **It moves food from the throat to the stomach.**

The walls of the esophagus are made of muscles. **These muscles move slowly to squeeze food so it passes through the body.** People do not usually feel their esophagus as it works.

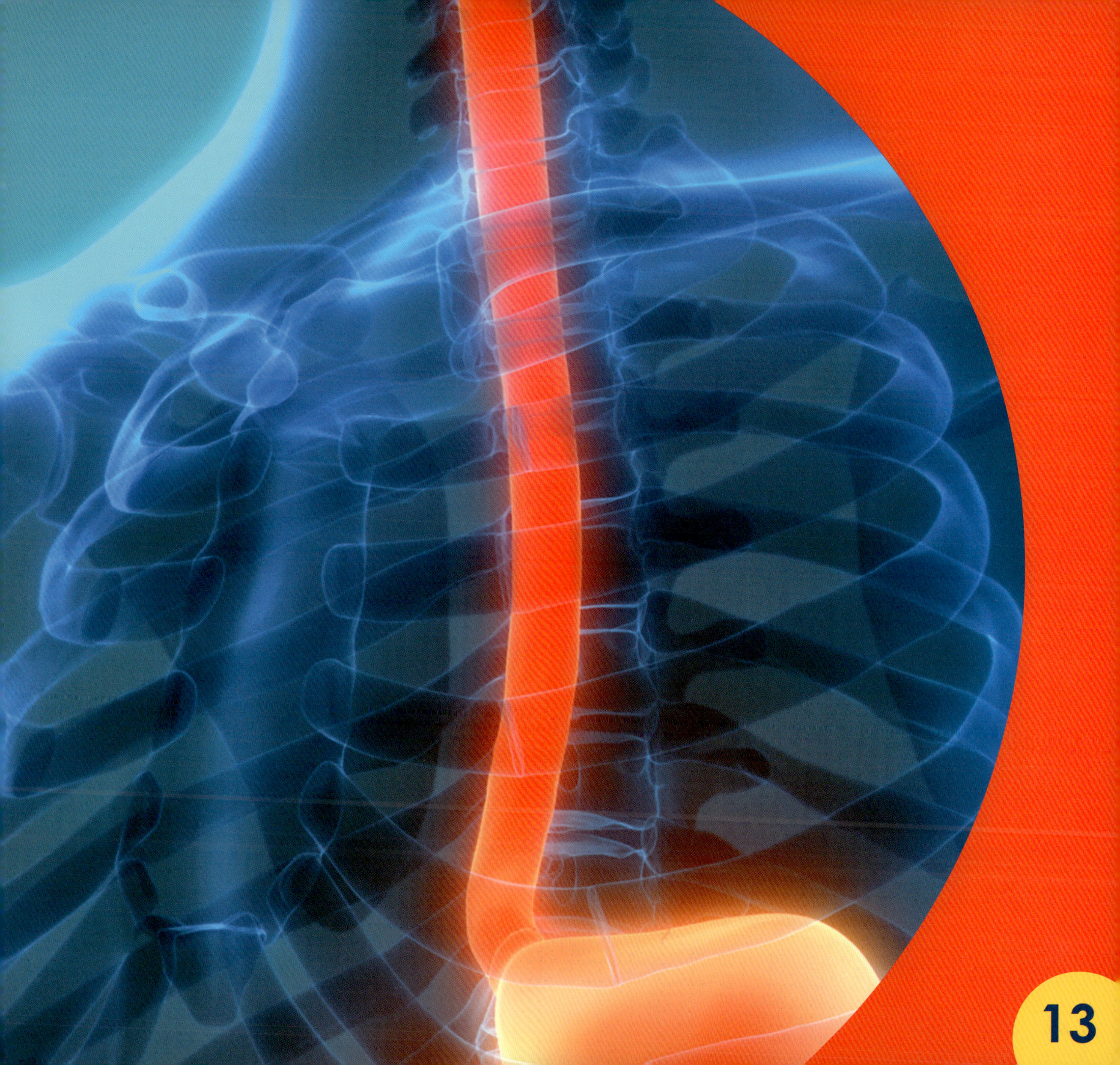

Stomach

The stomach is like a blender. **It breaks down and mixes the food that comes through the esophagus.** The stomach has strong muscles and digestive juices. These help it break food into smaller and smaller parts.

Food leaves the stomach as a thick mixture. **This goes into the small intestine.**

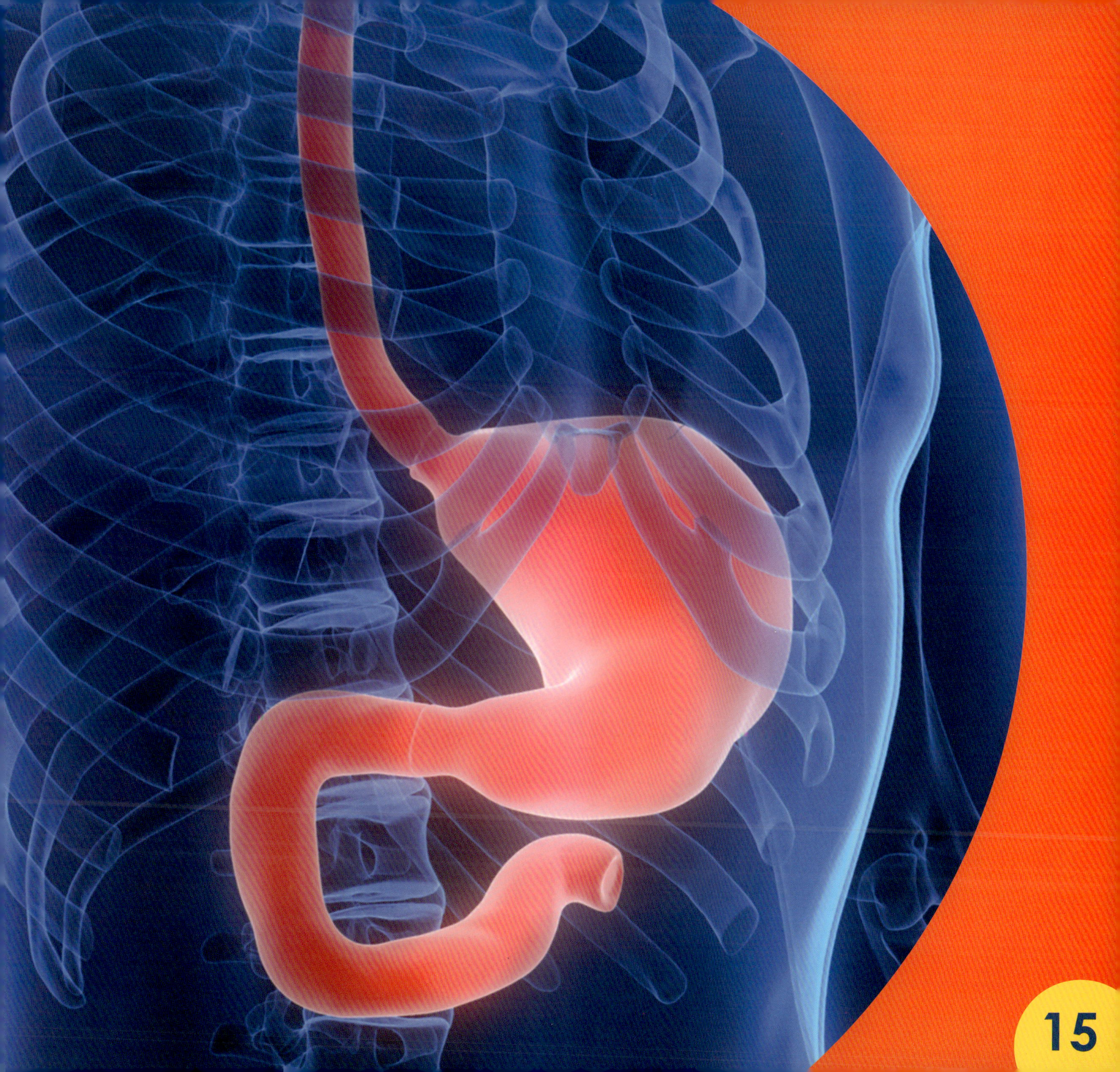

Intestines

The body has two types of intestines. They are the small and large intestines. **Food from the stomach enters the small intestine.** It is broken into even smaller parts.

The walls of the small intestine absorb nutrients from the food. Material that the small intestine cannot absorb moves to the large intestine. The large intestine changes the material into stool. This waste is stored in the rectum. When you go to the washroom, the stool is pushed out through the anus.

The small intestine absorbs about **95 percent** of the nutrients in food.

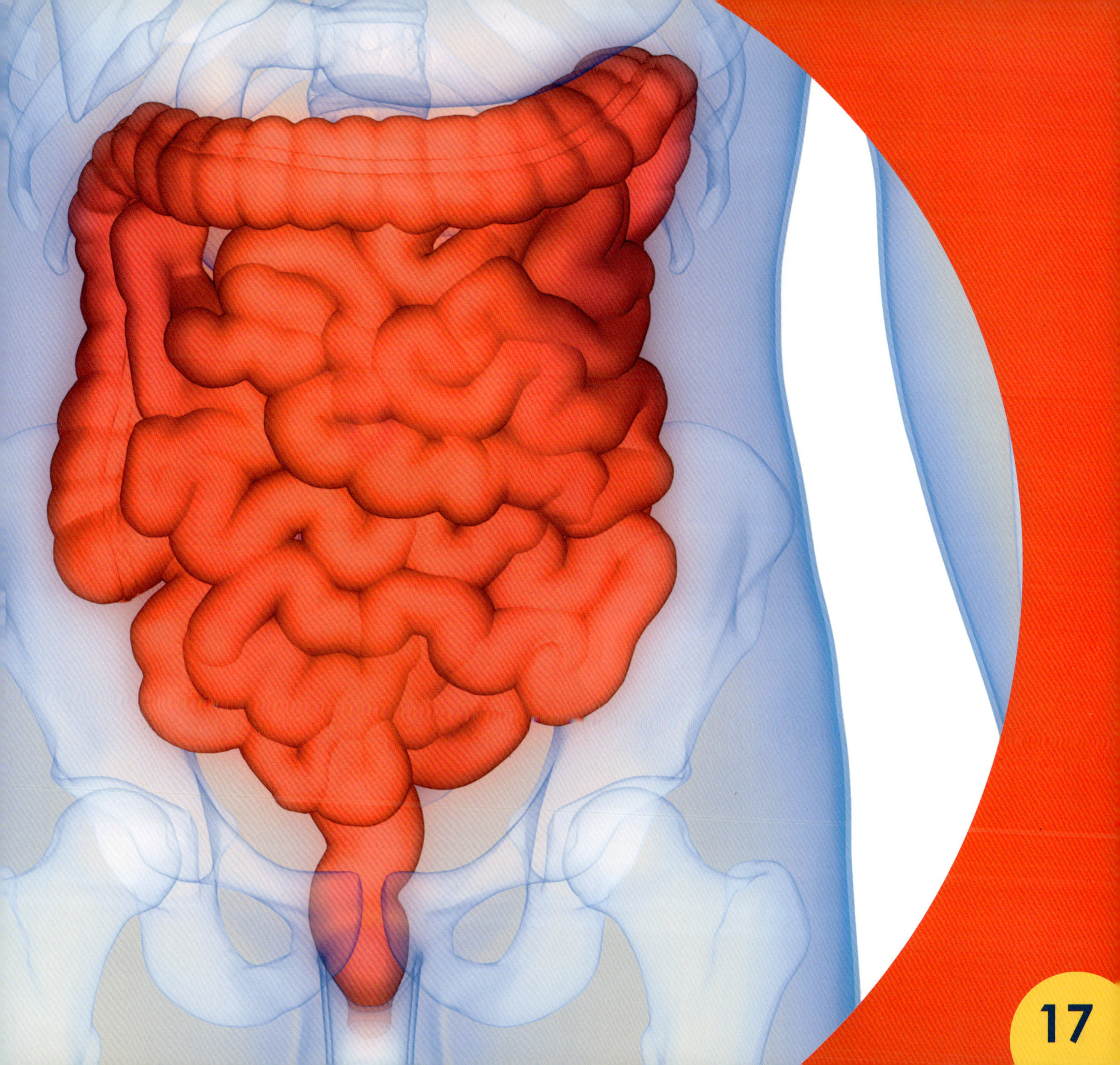

Staying Healthy

There are many foods that help the digestive system work well. **Fiber is very good for digestion.** Fruits, vegetables, and whole grains are often high in fiber.

Drinking plenty of water helps keep the digestive system healthy. Exercise and regular meals are also good for digestive health.

Career Spotlight

Gastroenterologists are doctors. They are experts on the digestive system. **Gastroenterologists help people with digestive problems.** They may take x-rays or do blood tests to find out what is wrong. Then, they make a plan to help the person feel better. The plan could include medicines and changes to exercise, food, or sleep.

Mayo Clinic in Minnesota is the best hospital in the United States for people with digestive problems.

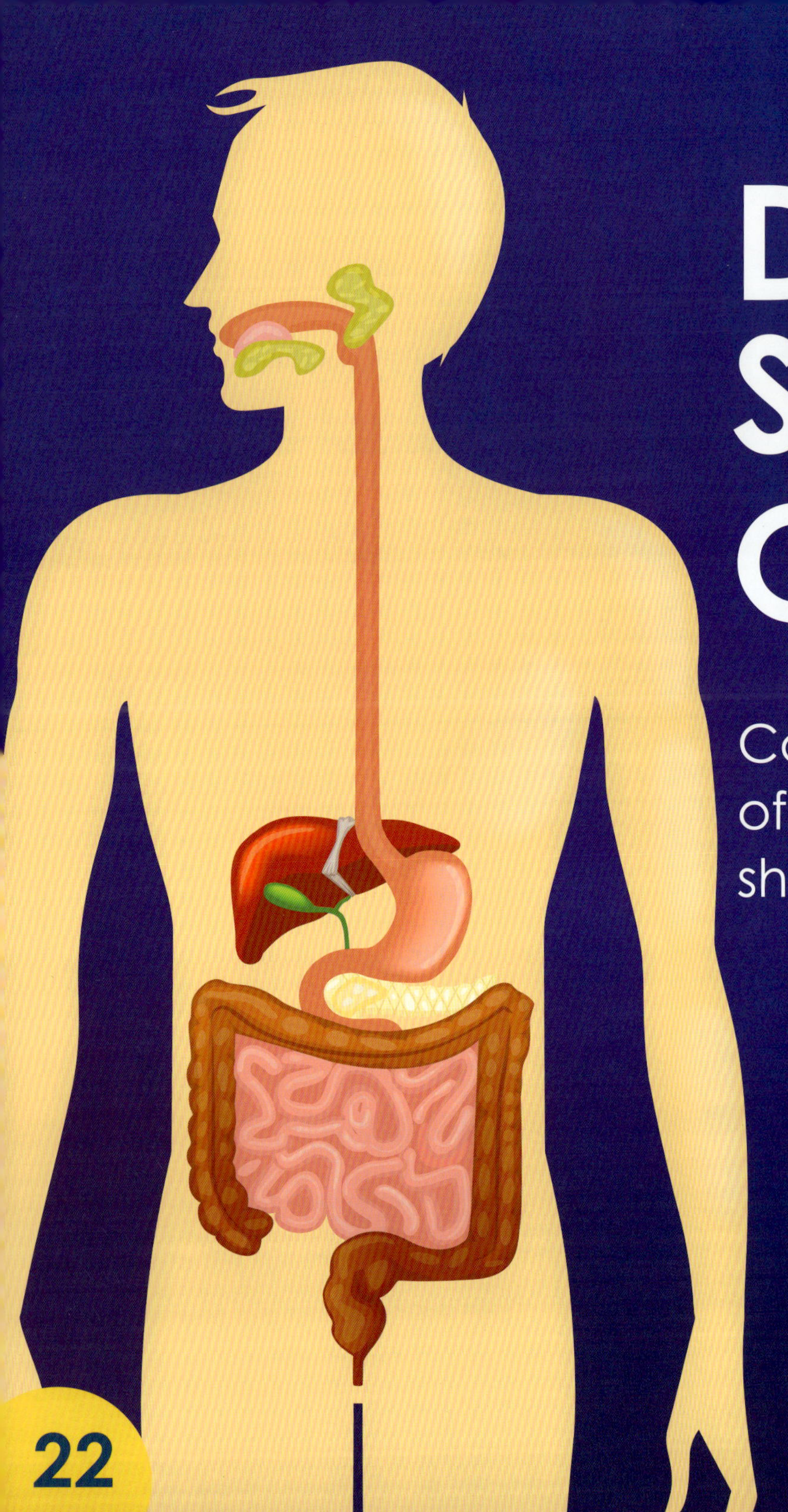

Digestive System Quiz

Can you name the parts of the digestive system shown in these pictures?

A] Mouth

B] Esophagus

C] Stomach

D] Intestines

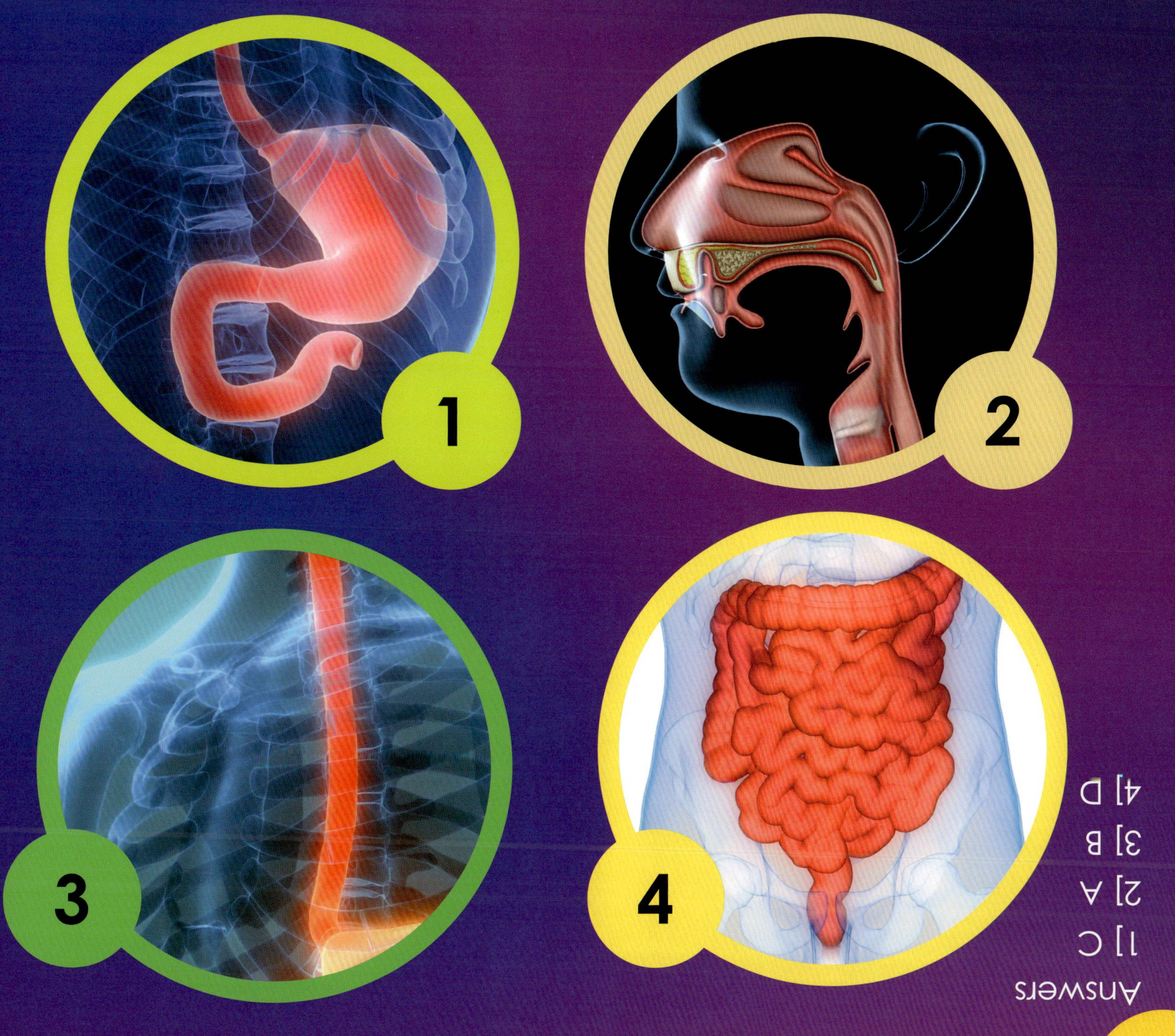

Answers
1] C
2] A
3] B
4] D

KEY WORDS

Research has shown that as much as 65 percent of all written material published in English is made up of 300 words. These 300 words cannot be taught using pictures or learned by sounding them out. They must be recognized by sight. This book contains 90 common sight words to help young readers improve their reading fluency and comprehension. This book also teaches young readers several important content words, such as proper nouns. These words are paired with pictures to aid in learning and improve understanding.

Page	Sight Words First Appearance
4	and, being, do, down, eat, every, food, for, from, gets, give, had, helps, in, is, it, needs, or, something, still, the, these, this, through, time, to, water, what, work, you, your
6	about, all, are, can, into, left, many, moves, of, over, part, small, some, that, there, use
8	a, long, runs, together, with
9	large
10	as, before, begins, even, have, makes, more, starts, take, their, then, when
12	also, made, not, people, so
14	comes, has, leaves, like
16	changes, go, out, they, two
18	good, high, often, very, well
19	keep
20	could, find, may, on, states

Page	Content Words First Appearance
4	body, breakfast, digestive system, energy, nutrients
5	absorption, digestion, functions, intestines, organs, stomach, waste
6	digestive tract
8	anus, mouth, tube
9	esophagus, rectum
10	adults, babies, bite, liquid, meal, process, saliva, teeth, tongue
12	muscles, pipe, throat, walls
14	blender, digestive juices, mixture
16	stool, washroom
18	apples, avocados, bananas, fiber, fruits, oats, spinach, vegetables, whole grains
19	exercise
20	blood tests, doctors, experts, gastroenterologists, hospital, Mayo Clinic, medicines, Minnesota, plan, problems, sleep, United States, x-rays

Published by AV2
14 Penn Plaza, 9th Floor New York, NY 10122
Website: www.av2books.com

Library of Congress Cataloging-in-Publication Data

Names: Erlic, Lily, author. | Das, Priyanka, author.
Title: Digestive system / Lily Erlic and Priyanka Das.
Description: New York, NY : AV2 by Weigl, [2021] | Series: My first look at body systems | Audience: Grades 2-3
Identifiers: LCCN 2020018412 (print) | LCCN 2020018413 (ebook) | ISBN 9781791118846 (library binding) | ISBN 9781791118853 (paperback) | ISBN 9781791118860 | ISBN 9781791118877
Subjects: LCSH: Digestive system--Juvenile literature.
Classification: LCC QP145 .E75 2021 (print) | LCC QP145 (ebook) | DDC 612.3--dc23
LC record available at https://lccn.loc.gov/2020018412
LC ebook record available at https://lccn.loc.gov/2020018413

Printed in Guangzhou, China
1 2 3 4 5 6 7 8 9 0 24 23 22 21 20

062020
100919

Project Coordinator: Priyanka Das Designer: Jean Faye Marie Rodriguez

Every reasonable effort has been made to trace ownership and to obtain permission to reprint copyright material. The publisher would be pleased to have any errors or omissions brought to its attention so that they may be corrected in subsequent printings.

The publisher acknowledges iStock and Shutterstock as the primary image suppliers for this title.